WORKBOOK

Biology

Higher tier

Suitable for GCSE Biology and GCSE Science Double Award

Answers online

James Napier

HODDER EDUCATION
LEARN MORE

Contents

(1) This workbook will help prepare you for the CCEA GCSE Biology and CCEA GCSE Double Award (Biology) exams.

(2) Your exams will include a range of short, structured questions and longer questions requiring continuous prose, including 6-mark questions testing both biological knowledge and your quality of written communication. You need to be able to answer questions requiring biological knowledge and understanding, as well as being able to analyse and evaluate data set in familiar and unfamiliar scenarios. This workbook will help you develop the skills to answer all these question types.

All questions are suitable for Higher Tier students. Questions in the paler tints are for Science Double Award and GCSE Biology. Questions in the darker tints are for GCSE Biology only.

(3) For each topic, there are:
- stimulus materials, including key terms and concepts
- short and longer exam-style questions
- space for you to write your answers

(4) Answering the questions will help you develop your skills and meet the assessment objectives AO1 (knowledge and understanding), AO2 (application) and AO3 (analysis and evaluation).

(5) You still need to read your textbook and refer to your revision guides and lesson notes.

(6) Timings are given for the exam-style questions to make your practice as realistic as possible.

(7) Marks available are indicated for all questions so that you can gauge the level of detail required in your answers.

(8) Answers are available at
www.hoddereducation.co.uk/workbookanswers

Cells

Cells and microscopy

The cell is the basic unit in living organisms. Plants, animals and bacteria are formed of different types of cell. Stem cells are special cells that can divide to form other types of cell as an organism grows. In multicellular organisms, the cells can be arranged into tissues, organs and organ systems.

1 The diagram below represents a bacterial cell.

Large loop of DNA

a i Identify the structure labelled X in the diagram. *1 mark*

....

ii State *one* structure in this cell that is also present in plant and animal cells. *1 mark*

....

iii State *two* differences between the structure of this cell and the structures of plant and animal cells. *2 marks*

1 ....

2 ....

b Explain how the use of electron microscopes has increased our understanding of cell structures. *2 marks*

....

2 a Complete the table below about some SI units that can be used in measuring biological structures. *3 marks*

Unit	Symbol	Number per metre in standard form
		1×10^3
Micrometre	μm	

b If the diagram of a bacterial cell at the start of this section represents a cell of 5 μm in length, calculate the magnification involved. *3 marks*

Show your working.

....

Stem cells, cell specialisation and diffusion

3 a Define the term 'stem cell'. *1 mark*

....

b Describe fully where stem cells occur in plants. *2 marks*

....

Questions marked **1 1** are for GCSE Biology students only

4 a Complete the table below about levels of organisation in organisms.

Level of organisation	Description
	Group of similar cells with same general structure and function
Organ system	
	Structure made of several types of tissue that carries out a particular function

b i Describe the effect of increasing size in living organisms on their surface area to volume ratio. **1 mark**

....

ii Explain the link between increasing size in living organisms and the need to have specialised exchange surfaces. **3 marks**

....

....

....

....

Exam-style questions ⑦

① Mitochondria are microscopic structures found in the cytoplasm.

a State the function of mitochondria. **1 mark**

....

Mitochondria produce carbon dioxide as a by-product. This carbon dioxide passes out of cells by diffusion.

b Explain why carbon dioxide diffuses out of cells. **2 marks**

....

....

c Name *two* parts of a cell that carbon dioxide diffuses through before leaving the cell. **2 marks**

1 ....

2 ....

d Apart from surface area, name *two* factors that will increase the rate of diffusion. **2 marks**

1 ....

2 ....

Questions marked ① ① are for Science Double Award and GCSE Biology students

⏱ 10

2 **a** Many patients who have leukaemia are given chemotherapy and/or radiotherapy as treatment. Once this treatment is complete they are often given a transfusion of stem cells. These stem cells often come from a close relative, such as a brother or sister, as their stem cells will be a good match.

   i Explain why a transfusion of stem cells is often necessary in patients with leukaemia. **2 marks**

....

....

   ii Name the type of cells these stem cells will produce. **1 mark**

....

   iii Suggest from which part of the body the stems cells are harvested in the donor individual. **1 mark**

....

   iv Suggest why it is important that the stem cells from the donor are a 'good match' to the tissue in the patient. **1 mark**

....

b Explain how embryonic stem cells differ from the type of stem cells used to treat leukaemia. **2 marks**

....

....

....

c A photograph of a stem cell in a textbook is 45 mm. It has been magnified 600 times. Calculate the actual length of the cell in μm. **3 marks**

Show your working.

.... μm

Photosynthesis and plants

Photosynthesis

Photosynthesis is a process in plants in which sugars and starches are built up from inorganic raw materials. The process requires light energy that is trapped by chlorophyll. Most photosynthesis takes place in the leaves of plants. Photosynthesis investigations often involve testing a leaf for starch to show that photosynthesis will only take place in the presence of chlorophyll, carbon dioxide and light.

1 a i State the word equation for photosynthesis.   *2 marks*

....

ii Complete the balanced chemical equation for photosynthesis below.   *2 marks*

$$6CO_2 + \text{.............................} \xrightarrow[\text{chlorophyll}]{\text{light}} \text{.............................} + 6O_2$$

b Photosynthesis is an endothermic process. In terms of photosynthesis, explain fully what is meant by the term 'endothermic'?   *2 marks*

....

....

....

Investigating photosynthesis

The starch test can be used to show that photosynthesis has taken place in a plant.

2 a Explain fully why plants are destarched in photosynthesis investigations.   *2 marks*

....

....

....

b Complete the sentences below.

When carrying out a starch test, leaves are boiled in .... to remove

the chlorophyll. This is done so that colour changes can be more easily seen when

.... is added to the leaf.   *2 marks*

3 a In what way are variegated leaves different from typical plant leaves?   *1 mark*

....

b Suggest why variegated leaves are not common in nature.   *1 mark*

....

Questions marked **1 1** are for Science Double Award and GCSE Biology students

Gas exchange and limiting factors

4 a i Name the process involving gas exchange that takes place in both light and darkness.

....

ii Photosynthesis also involves gas exchange. Name the gas taken into plant leaves during the process of photosynthesis.

....

iii Name the structures in plant leaves in which photosynthesis takes place. Circle the correct answer.

mitochondria   chloroplasts   nucleus   chlorophyll

b The table below shows how light intensity affects the rate of photosynthesis.

Light intensity/ arbitrary units	0	1	2	3	4	5	6	7	8	9	10
Rate of photosynthesis/ arbitrary units	0	1.5	3.1	4.5	6.1	7.6	7.6	7.5	7.6	7.7	7.6

i At which light intensity did light cease to become a limiting factor?

....

ii At a light intensity of 9 arbitrary units, state *two* environmental factors that could be limiting the rate of photosynthesis.

1 ....

2 ....

Leaf structure

5 a i Describe the function of guard cells and stomata.

....

....

ii In most plants, the stomata are mainly found on the lower surface of plant leaves. Suggest *two* reasons for this.

1 ....

....

2 ....

....

b Describe fully the function of the intercellular spaces in a leaf.

....

....

....

Exam-style questions

1 **a** Name a chemical used to absorb carbon dioxide. `1 mark`

....

b Describe how you could carry out an investigation to show that carbon dioxide is necessary for photosynthesis to take place. `5 marks`

....

....

....

....

....

....

....

2 The table below shows how hydrogencarbonate indicator changes colour in different carbon dioxide concentrations.

Carbon dioxide level	Low	Normal atmospheric level	High
Colour of bicarbonate indicator	Purple	Red	Yellow

a Describe what is meant by the term 'compensation point'. `1 mark`

....

....

b **i** You are provided with pondweed and normal laboratory apparatus. Describe how you could carry out an investigation to find the light intensity at the compensation point for the pondweed. `4 marks`

....

....

....

....

....

....

ii State *two* variables that you would need to control in this investigation. `2 marks`

1 ....

2 ....

Questions marked **1** **1** are for Science Double Award and GCSE Biology students

3 The diagram below summarises how the rate of photosynthesis changes from the top to the bottom of a typical leaf.

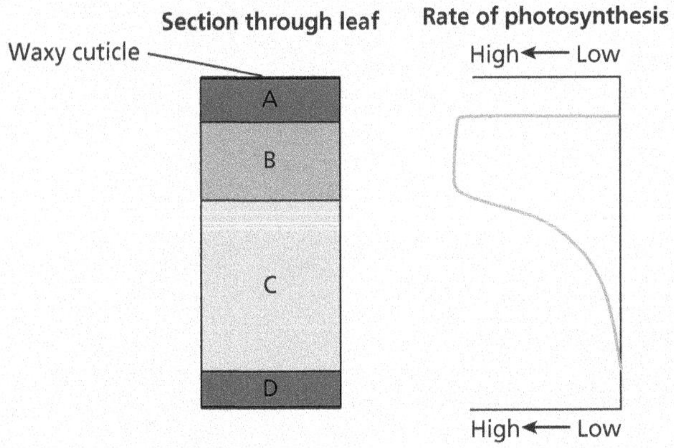

Section through leaf

Rate of photosynthesis
High ← Low

Waxy cuticle

A

B

C

D

High ← Low

The letters A–D refer to the different layers in a typical leaf.

a **i** Give *two* properties of the waxy cuticle in leaves. **2 marks**

1 ....

2 ....

ii State *three* ways in which the palisade mesophyll cells are adapted for photosynthesis. **3 marks**

1 ....

2 ....

3 ....

iii Identify the layers labelled A, B and C in the diagram. **2 marks**

....

....

b Describe and explain the changes in rate of photosynthesis across layers A, B and C. **4 marks**

....

....

....

....

....

....

....

....

....

Nutrition and food tests

Carbohydrates, fats and proteins are very important biological molecules. The presence of these molecules in food or any substance can be shown by using food tests. The energy content of different foods can be compared by burning food samples and measuring the temperature rise of the water.

Biological molecules

1 Carbohydrates contain either one or more basic sugar units.

    a   Place the carbohydrates below into order of number of units contained, starting with the smallest.

1 mark

        glycogen   glucose   lactose

....

    b   The diagram below represents part of a protein.

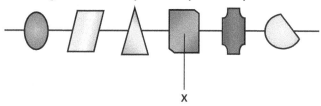

        X

       i   Identify structure X.

1 mark

....

       ii   Give *one* example of a functional protein in a cell.

1 mark

....

    c i   Name the component parts of a fat.

1 mark

....

       ii   State *one* function of fat in the body.

1 mark

....

2 Complete the table below about food tests.

3 marks

Food tested	Reagent	Initial colour	Final colour	Heating required?
Fat		Colourless	White emulsion	
	Benedict's		Brick-red precipitate	Yes
Starch		Yellow-brown		No

Questions marked 1 1 are for Science Double Award and GCSE Biology students

Exam-style questions

1 a Name the reagent used to test for protein.

....

b Proteins are made up of amino acids bonded together. The human body can contain up to 100 000 different proteins and yet there are only 20 different types of amino acids.

Suggest how many different types of protein can be formed by as few as 20 different types of amino acid. **2 marks**

....

....

....

2 Streaky bacon contains large quantities of protein and fat. The energy content of two pieces of streaky bacon (**A** and **B**) was compared by burning a small piece of each under a boiling tube of water and measuring the temperature increase of the water for each piece of bacon. The results of the investigation are shown below.

Sample	Initial temperature of water/°C	Final temperature of water/°C	Temperature increase/°C
A	16	36	20
B	16	40	24

a The percentage increase in temperature of the water in sample **A** was 125%. Calculate the percentage increase for sample **B**. **2 marks**

Show your working.

.... %

b State *three* variables that would have been controlled to ensure the investigation was valid. **3 marks**

1 ....

2 ....

3 ....

c Apart from experimental error, use the information provided to suggest *one* reason why the temperature increase in the two bacon samples was different. **1 mark**

....

....

Questions marked **1** **1** are for GCSE Biology students only

Enzymes and digestion

Enzymes are biological catalysts that speed up reactions without being used up themselves. Enzyme action can be affected by temperature, pH, enzyme concentration and inhibitors. Enzymes are very important in the digestive system where they break down large insoluble molecules into small soluble molecules that can be absorbed. Most absorption in the human body takes place in the ileum, an organ that is highly adapted for this function.

Enzymes

1 a Name the biological molecule from which enzymes are made.

1 mark

....

b Complete the sentences below.

3 marks

Substrate molecules are complementary in shape to the .... of an

enzyme, and an enzyme will only produce a reaction if the substrate is an exact fit.

This is described as the .... model of enzyme action. This model

explains the idea of substrate .... , which is the idea that each

enzyme can only catalyse one (or a small range of) substrates.

2 Complete the table below about different types of enzyme and their functions.

3 marks

Enzyme	Substrate	Products(s)
	Starch	Sugar
Protease	Protein	
Lipase	Lipids (fats)	

3 The graph below shows the effect of temperature on enzyme activity.

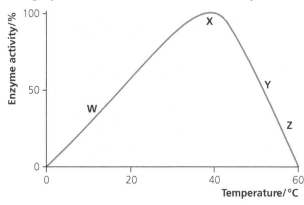

a State the term used to describe the temperature at which maximum enzyme activity takes place (shown as X on the graph).

1 mark

....

Questions marked 1 1 are for Science Double Award and GCSE Biology students

b Complete the table below.

Position on graph	Rate of enzyme activity	Explanation
W	Low	
Y	Moderate	
Z	Very low	

Digestion and absorption

4 Apart from a large surface area, state *two* ways in which the ileum is adapted for absorption. **2 marks**

1..

2..

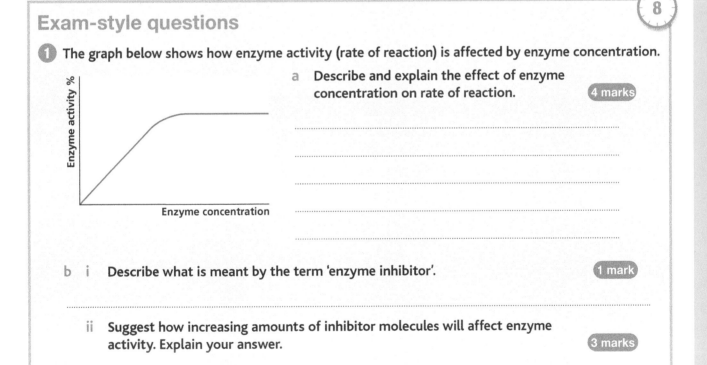

Exam-style questions

1 The graph below shows how enzyme activity (rate of reaction) is affected by enzyme concentration.

a Describe and explain the effect of enzyme concentration on rate of reaction. **4 marks**

....

....

....

....

....

b i Describe what is meant by the term 'enzyme inhibitor'. **1 mark**

....

ii Suggest how increasing amounts of inhibitor molecules will affect enzyme activity. Explain your answer. **3 marks**

....

....

....

....

2 **a** The diagram below shows a section through part of the wall of the ileum.

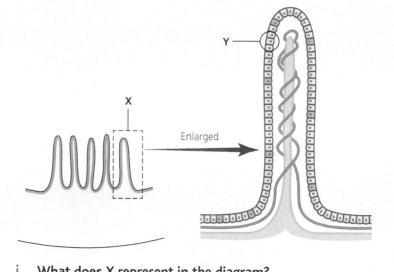

Y

X

Enlarged

    i What does X represent in the diagram? `1 mark`

    ii On the diagram, label the lacteal and the capillary network. `2 marks`

    iii Describe the function of the lacteal. `1 mark`

    iv Explain how layer Y is adapted for absorption. `2 marks`

    v Villi on the wall of the ileum increase the surface area across which food can be absorbed. State *two* other features of the ileum that increase surface area for absorption. `2 marks`

1

2

  b In people with coeliac disease, the villi become flattened or destroyed. Suggest how this will affect the normal functioning of the ileum. Explain your answer. `2 marks`

Questions marked ① ① are for Science Double Award and GCSE Biology students

The respiratory system, breathing and respiration

Respiration is the process that releases energy in all cells. Mitochondria are the sites of cellular respiration. The respiratory system ensures that respiratory gases get to and from the respiratory (exchange) surfaces in plants and animals. Respiration can take place in the presence of oxygen (aerobic) or in the absence of oxygen (anaerobic).

The respiratory system and lung model

1 **a** **Complete the table below about some structures in the respiratory system.** **4 marks**

Structure	Function
Alveoli	
	The structures that link alveoli to the bronchi
Intercostal muscles	
Pleural fluid	

b **The following statements describe breathing in (inhalation) in humans but they are not in the correct sequence.**

1 There is a reduction of air pressure in the thoracic cavity.

2 The volume of the thoracic cavity increases.

3 Air enters the lungs.

4 The ribs (chest wall) move out and the diaphragm flattens.

Place the statement numbers in the correct sequence to describe the process of inhalation. **1 mark**

....

2 **Respiratory surfaces in plants and animals have several features in common. State *three* of these features.**

1 ....

2 ....

3 ....

Respiration

3 a Write the word equation for aerobic respiration.

2 marks

....

b The table below compares aerobic and anaerobic respiration in humans. Complete the table by striking out the incorrect alternatives (the first row has already been completed).

3 marks

Feature	Type of respiration	
	Aerobic	Anaerobic
Oxygen used	Yes/~~No~~	~~Yes~~/No
Carbon dioxide produced	Yes/No	Yes/No
Lactic acid produced	Yes/No	Yes/No
Amount of energy produced	High/Low	High/Low

c State *one* difference between anaerobic respiration in yeast and anaerobic respiration in mammalian muscle.

1 mark

....

Exam-style questions

(12)

1 The diagram below represents a model of the respiratory system.

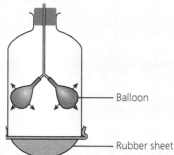

Balloon

Rubber sheet

a i Describe the sequence of events that causes the balloons to inflate.

3 marks

....

....

....

....

ii Name the part of breathing represented by the model.

1 mark

....

b The bell-jar model shown does not fully represent how breathing takes place in humans. State *two* ways in which the model fails to represent how breathing takes place in humans.

2 marks

1 ....

....

2 ....

....

Questions marked **1** **1** are for Science Double Award and GCSE Biology students

c The effect of exercise on breathing is shown in the graph below.

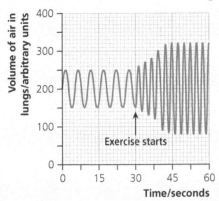

i Calculate the difference between the maximum and minimum volumes of air in the lungs before exercise.

Show your working.

.... arbitrary units

ii During exercise the depth of breathing increased and levelled off. State *one* other change in breathing that took place during exercise. **1 mark**

....

iii Explain the reason for the changes in breathing during exercise. **3 marks**

....

....

....

④

2 The graph below shows changes in a flask containing yeast in a glucose solution.

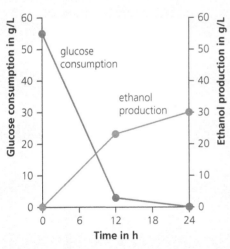

a Name the process taking place in the flask. **1 mark**

....

b Suggest why the glucose consumption and ethanol production levelled off in the second half of the graph.

....

....

....

c Apart from changes in glucose concentration and ethanol concentration, suggest *one* other change that takes place in the flask over the 24 hours. **1 mark**

....

The nervous system and hormones

The nervous system and hormones enable animals to be aware of their environment (both external and internal) and allow them to respond to changes. Osmoregulation involves the regulation of water balance in the body and this is an example of homeostasis — the process of maintaining a constant internal environment. Diabetes is an example of a condition caused by the process of homeostasis breaking down and failing to work effectively.

The eye

1 **a** Name *two* parts of the eye that refract light. 2 marks

1 ....

2 ....

b Name the part of the eye that controls the amount of light entering the eye. 1 mark

....

c The process of accommodation explains how the eye can focus on both near and distant objects. This is achieved by the shape (thickness) of the lens changing. Describe fully the changes within the eye when focusing on a distant object. 3 marks

....

....

Neurones and synapses

2 The diagram below represents a neurone.

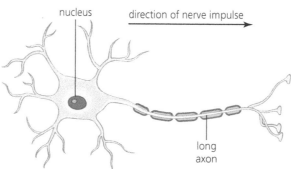

a Label the myelin sheath on the diagram. 1 mark

b Apart from the presence of a myelin sheath and a cell body, state *two* other ways in which this neurone is adapted for its function. 2 marks

1 ....

2 ....

c Circle *one* part of the neurone that would produce transmitter chemical. 1 mark

Questions marked **1** **1** are for Science Double Award and GCSE Biology students

Voluntary and reflex actions

3 **a** **i** Complete the sentences below.

The overall length of the neurones involved in reflex actions is shorter than that for

.... actions. This enables reflex actions to be.., so

reflex actions often have a protective role.

ii Give *one* example of a reflex action.

b The following statements describe the structures involved in a reflex action, but
they are not in the correct order.

A effector

B motor neurone

C sensory neurone

D receptor

E association neurone

Place the letters in the correct order to show the sequence of the structures
involved in a reflex arc.

....

Hormones and diabetes

4 **a** Define the term 'hormone'.

b **i** Name the organ that produces insulin.

ii What will cause this organ to increase insulin production?

iii Describe fully the liver's role in insulin action.

5 **a** Define the condition diabetes.

b Explain the difference between a symptom and a long-term effect (complication)
of diabetes.

Questions marked ⓫ ⓫ are for GCSE Biology students only

ii The descriptions below are either symptoms or long-term effects of diabetes.

kidney failure   lethargy   eye damage   glucose in the urine

Place each description in the correct column of the table below. 2 marks

Symptoms	Long-term effects

Excretory system and osmoregulation

6 a i Name the components of the excretory system, in order, through which urine passes from leaving a kidney to exiting the body. 1 mark

.... → .... → ....

ii State *three* ways in which the body loses water. 3 marks

1 ....

2 ....

3 ....

b Describe the role of the kidney in osmoregulation. 3 marks

....

....

....

....

Plant hormones

7 a Define the term 'phototropism'. 1 mark

....

....

b Describe fully the action of auxin in phototropism. 4 marks

....

....

....

....

....

Questions marked 🔵 🔵 are for Science Double Award and GCSE Biology students

Exam-style questions

1 **a** The diagram below represents a synapse.

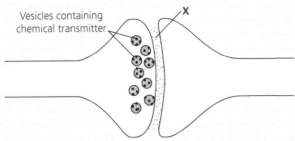

Vesicles containing
chemical transmitter

X

     i Add an arrow to the diagram to show the direction nerve impulses travel. `1 mark`

     ii Describe how a nerve impulse can pass from one neurone to an adjacent neurone at a synapse. `4 marks`

....

....

....

....

     iii Why is it important that the gap between adjacent neurones (X in the diagram) is very short? `2 marks`

....

....

b Many reflex actions, such as blinking and the withdrawal reflex, are protective. Reflex arcs usually have fewer synapses than voluntary nerve pathways between the same receptors and effectors. Suggest the advantage of this. `2 marks`

....

....

2 Insulin is the hormone used in the control of blood glucose. When blood glucose levels are high, more insulin is released. As blood glucose levels return to normal, the amount of insulin produced drops.

    a State the term that describes this reduction in level of insulin produced as glucose levels return to normal. `1 mark`

....

    b In terms of *initial* treatment, give *one* difference between type 1 diabetes and type 2 diabetes. `1 mark`

....

    c The table below shows the average age when patients were diagnosed with type 2 diabetes in a hospital over a 15-year period.

Year	Average age when diagnosed
1990	61
1995	58
2000	54
2005	51

Questions marked **1** **1** are for GCSE Biology students only

   i   State the trend shown by the data in the table. (1 mark)

....

   ii Suggest *two* reasons that could account for this trend. (2 marks)

1 ....

2 ....

   iii Scientists wanted to investigate trends in type 2 diabetes using data from this hospital. Suggest *one* other piece of information, apart from average age at diagnosis, they would require. (1 mark)

....

d Diabetes is a condition that costs the NHS a lot of money — up to 10% of its total budget. Many people have diabetes and the instruments that monitor blood glucose levels are expensive.

Doctors, nurses and other specialists place great emphasis on the value of educating people with diabetes about the importance of monitoring regularly and controlling their blood glucose concentrations. Having diabetes for a long time and/or poor control of blood glucose concentration can lead to long-term effects (complications) of diabetes.

   i   Using only the information provided, state *three* reasons why the cost to the NHS of treating people with diabetes is so high. (3 marks)

1 ....

2 ....

3 ....

   ii Explain fully the benefits of educating people with diabetes about the importance of having good blood glucose concentration control. (2 marks)

....

....

....

3 The kidney is the organ responsible for osmoregulation.

a Describe what is meant by the term 'osmoregulation'. (1 mark)

....

b Name the blood vessel that carries blood to the kidney. (1 mark)

....

c The table below shows the amount of antidiuretic hormone (ADH) produced per hour by a student after carrying out a period of exercise on a hot day. Throughout a normal day, this student produces, on average, around 2 arbitrary units of ADH per hour.

Time/hours after exercise	1	2	3	4	5	6	7	8
ADH production/arbitrary units per hour	10	9	10	6	4	2	2	2

i   Describe the results shown. Use data from the table to support your answer.   **3 marks**

....

....

....

....

ii Using the information provided, explain how and why the amount of ADH produced
    would have affected the volume of urine produced in the first 3 hours.   **4 marks**

....

....

....

....

Ecological relationships and energy flow

Ecological relationships describe how living organisms interact with each other and the environment. Food chains and food webs describe how the sun's energy is utilised by green plants and then used by other living organisms. The carbon and nitrogen cycles show how elements are recycled in nature. Global warming and eutrophication are examples of how people's poor management of the environment has caused harm to the natural world and biodiversity.

Biological terms and fieldwork

1 a **Define the terms 'population' and 'ecosystem'.** 2 marks

Population

....

Ecosystem

....

b **Factors that affect living organisms can be abiotic or biotic. Circle the biotic factors in the row below.** 2 marks

pH   light   competition   temperature   predation

2 a **If a square quadrat has an area of $0.25\,m^2$, what is the length of each side?** 1 mark

....

b i **Quadrats are used to sample habitats. Describe how you would use a quadrat in a random sampling exercise to estimate the number of plantain plants in a school football pitch.** 4 marks

....

....

....

....

....

....

ii **If the football pitch was $100\,m \times 50\,m$ and the average number of plantain plants per $0.25\,m^2$ quadrat was 5, calculate the estimated number of plantain plants in the football pitch.** 3 marks

Show your working.

.... **plants**

Food chains, food webs and energy flow

3 a The diagram below shows a food chain.

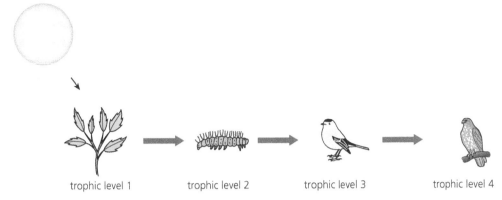

trophic level 1   trophic level 2   trophic level 3   trophic level 4

    i Which trophic level represents a producer? *1 mark*

....

    ii How many consumers are there in total? *1 mark*

....

b Sketch a pyramid of biomass for this food chain. **3 marks**

c Explain fully why short food chains are more efficient than longer food chains. **2 marks**

....

....

Decomposition, the carbon cycle and global warming

4 a i Define the term 'decomposition'. *1 mark*

....

    ii Saprophytic fungi are decomposers. Name *one* other group of living organisms
that carries out decomposition. *1 mark*

....

b Describe the process of nutrition in saprophytic fungi.

....

....

....

5 The diagram below summarises the carbon cycle.

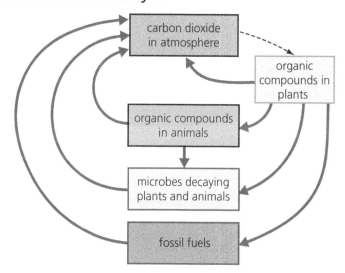

Use the diagram and your understanding to answer the questions below.

a Name the only process that takes carbon dioxide out of the atmosphere. 1 mark

....

b Place an R on the arrow(s) that represent respiration taking place. 1 mark

c Place a C on the arrow(s) that represent the process of combustion. 1 mark

Minerals and the nitrogen cycle

6 a Complete the table below about minerals and their functions. 2 marks

Mineral	Function
Magnesium	
	For making cell walls

b Minerals can be added to the soil in the form of natural or artificial fertiliser.

i Name *one* natural fertiliser. 1 mark

....

ii Give *two* advantages of using natural fertiliser compared with artificial fertiliser. 2 marks

1 ....

2 ....

7 a i   Describe precisely where root hair cells are located.

ii   Draw a root hair cell in the space below. In your drawing, you should label the
cell wall, nucleus, cytoplasm and vacuole.

b   Minerals can be taken into plant roots by active transport. Describe the process of
active transport.

Eutrophication and human activity and biodiversity

8 a i   In the process of eutrophication, name the mineral that stimulates the growth
of algae and aquatic plants.

ii   State *one* thing that can cause high levels of this mineral to enter waterways.

Over time, many of the numerous aquatic plants and algae growing in the waterway as a
consequence of eutrophication die.

b   Explain how the death of these plants can eventually lead to the death of animal
life in the waterway.

9   The process of reforestation can help promote biodiversity and help the environment in
a number of ways.

a   Define the term 'reforestation'.   1 mark

b i   Give *two* ways in which reforestation can help promote biodiversity.   2 marks

1

2

ii   Explain fully how reforestation can help the environment.   2 marks

Exam-style questions

1 Some students wished to investigate the distribution of a seaweed on a rocky shore. They placed a tape running from the low tide mark to the high tide mark up the shore. The students placed a 1 m² quadrat each metre along this tape starting from the low tide mark.

a Name the type of transect used.

1 mark

....

b The students used percentage cover to record the distribution of the seaweed. Their results are shown in the graph below.

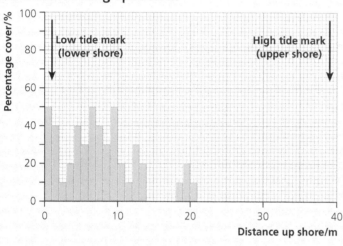

i What percentage of quadrats had this seaweed present?

Show your working.

2 marks

.... %

ii Describe fully the distribution of the seaweed on the shore. Use data from the graph to support your results.

3 marks

....

....

....

iii In terms of adaptation to its environment, suggest *one* way in which this seaweed is adapted.

1 mark

....

c In grassland habitats, soil pH is an abiotic factor that can affect the distribution of plants that will grow.

i What apparatus would you use to measure the pH of soil?

1 mark

....

ii State *one* other abiotic factor that you could measure when doing fieldwork. What apparatus would you use?

2 marks

....

Questions marked **1** **1** are for Science Double Award and GCSE Biology students

2 **a** Name *two* resources that plants compete for.　　　　　　　　　　　2 marks ⑨

1 ....

2 ....

b Rhododendron is a large, rapidly spreading, evergreen shrub with dense, large leaves that prevent much light from passing through to the ground. It was introduced by people to Britain a few centuries ago and has spread rapidly through many areas. In areas where there is a lot of rhododendron, biodiversity (the number of species of plants and animals) is greatly reduced.

　　i Using the information provided, explain fully why biodiversity is reduced in areas where there is a lot of rhododendron.　　　　　　　5 marks

....

....

....

....

....

　　ii Conservation volunteers and other groups are working to reduce the effect that rhododendron has on the environment. Suggest *one* way they can reduce the impact of the rhododendron.　　　　　　　1 mark

....

c Animals also compete for resources. Name *one* resource that only animals compete for (but not plants).　　　　　　　1 mark

....

3 **a** Garden compost was placed in a large lidded bin. Holes were drilled in the side of the bin to help speed up the composting process (decomposition). The table below shows how the temperature at the centre of a compost heap changed over a 20-week period.　　⑰

Week	1	3	5	7	9	11	13	15	17	19	21
Temperature/°C	20	21	23	27	32	38	48	42	37	33	24

　　i Describe the trend shown by this information. Use data from the table to support your answer.　　　　　　　3 marks

....

....

　　ii Explain the change in temperature in the compost bin.　　　　　　2 marks

....

　　iii Suggest why holes were drilled in the side of the bin.　　　　　　3 marks

....

....

b   Suggest why the formation of compost takes place much faster in the summer than in the winter.
    (1 mark)

....

c   Many plants such as clover contain root nodules. These root nodules contain nitrogen-fixing bacteria that convert nitrogen gas into nitrates. Scientists investigated the number of root nodules in clover plants in a series of fields that had different amounts of fertiliser added. The results are shown in the table below.

Sample	1	2	3	4	5	6	7	8	9	10
Fertiliser concentration/arbitrary units	0	2	3	4	5	6	7	8	9	10
Number of root nodules	25	23	20	21	17	15	13	13	10	8

    i   Describe the results shown by the table.
        (1 mark)

....

    ii Suggest an explanation for these results.
        (3 marks)

....

....

....

d   The scientists also compared the relative amounts of nitrate in a number of fields. They found that waterlogged fields usually had lower levels of nitrate than fields with good drainage. Use your knowledge of the nitrogen cycle to explain this difference in nitrate levels.
    (4 marks)

....

....

....

....

....

4   Describe the process of global warming. Your answer should include the causes of global warming and the problems associated with it.

    In this question you will be assessed on your written communication skills, including the use of specialist scientific terms.
    (6 marks)

....

....

....

....

....

....

....

....

....

....

....

Questions marked (1) (1) are for Science Double Award and GCSE Biology students

Osmosis and plant transport

Osmosis is the term that describes the movement of water into and out of cells. Osmosis can explain how plant cells become turgid or plasmolysed. Transpiration describes the loss of water by evaporation from plant leaves. The rate of transpiration in plants is affected by wind speed, temperature, humidity, light and leaf surface area.

Osmosis

1 a **Define the term 'osmosis'.**

....

....

....

b **The following statements describe how a plant cell X becomes turgid but they are not in the correct sequence.**

1 **As a result, water enters X by osmosis.**

2 **Water enters X until the cell membrane is pushed tightly against the cell wall.**

3 **The cell contents of X are more concentrated than the surrounding cells.**

Place the statement numbers in the correct sequence to describe the process of turgor in plant cells.

....

2 a **A potato cylinder was weighed and placed in a concentration of 15% sucrose. 24 hours later it was reweighed and found to have a reduced mass. Explain fully this result.**

....

....

....

b **In investigations involving osmosis, Visking tubing is often used. Name the part of a cell that the tubing represents.**

....

Transpiration

3 a **Complete the sentence below to describe the process of transpiration.**

Transpiration is the evaporation of water from .... cells, followed by

diffusion through .... and then the ....

b **Describe and explain the effect of temperature on the process of transpiration.**

....

....

....

Questions marked **1** **1** are for GCSE Biology students only

4 a The diagram below shows a potometer.

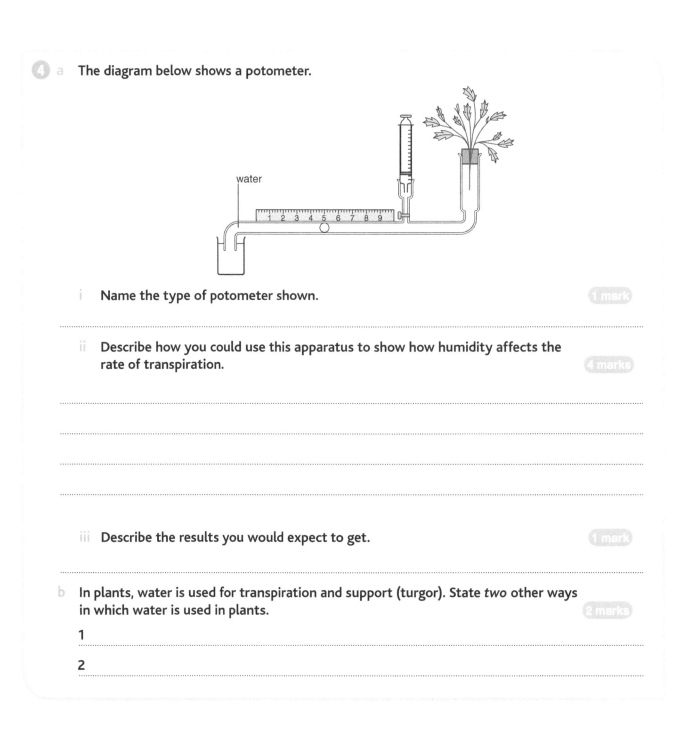

water

1 2 3 4 5 6 7 8 9

i Name the type of potometer shown.

1 mark

....

ii Describe how you could use this apparatus to show how humidity affects the rate of transpiration.

4 marks

....

....

....

....

iii Describe the results you would expect to get.

1 mark

....

b In plants, water is used for transpiration and support (turgor). State *two* other ways in which water is used in plants.

2 marks

1 ....

2 ....

Exam-style questions

1 **a** Five potato cylinders were cut to approximately 60 mm in length, which was measured as accurately as possible. They were then placed in beakers of water, 2.5% sucrose, 5% sucrose, 7.5% sucrose and 10% sucrose respectively. After 2 hours, the potato cylinders were removed from the beakers and their lengths measured. The results are shown in the table below.

| Solution/% sucrose | Length of potato cylinder | | | |
	Initial length/ mm	Final length/ mm	Change/mm	Percentage change/%
0 (water)	61	64	+3	+4.9
2.5	60	62	+2	+3.3
5.0	59	60	+1	+1.7
7.5	61	61	0	0
10.0	61	60	−1	

i Complete the table by calculating the percentage change for the potato in the 10% sucrose.

Show your working.

 2 marks

.... %

ii State the concentration of sucrose that has the same concentration as the potato. 1 mark

b The investigation was repeated using sweet potato. In water, the percentage increase was 10.0%; in 2.5% sucrose 8.1%; in 5% sucrose 5.9%; and in 7.5% sucrose 4.2%. In 10% sucrose there was a 1.9% increase.

i Estimate the sucrose concentration that would result in no percentage change in length. 1 mark

ii Suggest an explanation for the difference between these results and the results for the potato. 2 marks

iii State *two* variables that would need to have been controlled when comparing the results for the potato and the sweet potato. 2 marks

1 ....

2 ....

2 **a** The graph below shows how wind speed affected the rate of transpiration in a particular plant during the day and during the night.

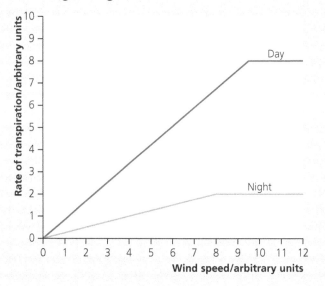

i Describe how wind speed affected the rate of transpiration during the *day*. Use data from the graph to support your answer.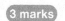

....

....

....

....

ii Explain the effect of wind speed on transpiration during the day.

....

....

....

iii Explain the change of transpiration rate at night compared with during the day.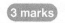

....

....

b **i** Leaf surface area also affects transpiration rate. Describe and explain the relationship between leaf surface area and rate of transpiration.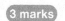

....

....

....

ii In many plants, the stomata close in very hot weather. Give *one* advantage and *one* disadvantage of this to the plant.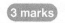

Advantage ....

Disadvantage ....

Questions marked **1** **1** are for Science Double Award and GCSE Biology students

The circulatory system

The circulatory system is a transport system in animals that carries materials throughout the body. The transport medium is blood, which is carried in a continuous loop within specialised blood vessels called arteries, veins and capillaries. The heart is the organ that provides the force that pushes the blood around the body.

Blood and blood vessels

1 a **Using lines, link the blood component to its most appropriate function.**

Blood component	Function
Red blood cell	Defence against disease
White blood cell	Transports carbon dioxide
	Transports oxygen

b **Name the type of blood cell that does not contain a nucleus.**

....

2 a **Complete the table below, showing some features of the main types of blood vessel.**

Feature	Blood vessel		
	Artery	Vein	Capillary
Walls are thick (relative to lumen)	Yes		No
Presence of valves	No	Yes	
Carries blood under high pressure	Yes		No

b i **Name the only artery that carries deoxygenated blood.**

....

ii **Name the type of blood vessel that contains the most muscle.**

....

3 a **Name the organs and blood vessels, in sequence, that the blood flows through from leaving the heart until reaching the renal vein.**

....

b **State *three* ways in which the blood in the renal vein will be different from the blood that has just left the heart.**

1 ....

2 ....

3 ....

The heart

4 a The diagram below shows the structure of the heart.

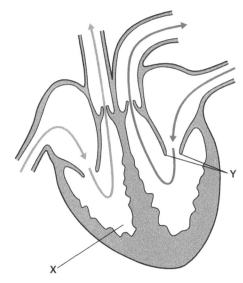

   i Name the part of the heart labelled X. *1 mark*

....

   ii On the diagram, label the pulmonary vein. *1 mark*

   iii State the function of the structures labelled Y. *2 marks*

....

....

b Explain why the walls of the ventricles are thicker than the walls of the atria. *2 marks*

....

....

5 a i Describe what is meant by the term 'cardiac output'. *1 mark*

....

   ii Describe the effect that exercise has on cardiac output. *1 mark*

....

b State *one* other effect regular exercise has on the heart. *1 mark*

....

Exam-style questions

1 a Red blood cells are highly adapted for the function of transporting oxygen. State *two* adaptations and, for each, explain how the adaptation helps to maximise the amount of oxygen carried in each red blood cell. **4 marks**

1 ....

....

2 ....

....

b During exercise, the pulse rate increases in response to increased rates of contraction of the muscles involved in the exercise. Explain why the pulse rate increases and explain the advantage of this to the muscles involved in exercise. **4 marks**

....

....

....

....

c The concentration of blood plasma in humans is kept at a constant level. Describe and explain the effect that a more dilute blood concentration would have on red blood cells. **2 marks**

....

....

....

10

2 Arteries, veins and capillaries are different types of blood vessel found in the circulatory system. Describe how the structure of each of these blood vessels is linked to its function.

In this question you will be assessed on your written communication skills, including the use of specialist scientific terms. **6 marks**

....

....

....

....

....

....

....

....

....

....

....

....

6

Questions marked **1** **1** are for GCSE Biology students only

Reproduction, fertility and contraception

The male reproductive system is adapted for producing sperm and getting the sperm into the female reproductive system. The female system is adapted to produce eggs and to nourish and protect the embryo/foetus should pregnancy occur. Pregnancy may fail to occur due to the male and/or the female being infertile, or through the use of contraception.

The male and female reproductive systems

1 a **The table below provides some information about the male reproductive system. Complete the table.** *3 marks*

Structure	Function
Testis	
	Nourishes sperm
Sperm tubes	

b **Sperm cells are adapted by having a haploid nucleus and mitochondria for energy production.**

  i **Explain fully why sperm cells have a high energy requirement.** *2 marks*

  ....

  ....

  ii **State *one* other way in which sperm cells are adapted for their function.** *1 mark*

  ....

  iii **In which part of the female reproductive system does fertilisation take place?** *1 mark*

  ....

Infertility

2 a **Give *two* causes of infertility in human females.** *2 marks*

  1 ....

  2 ....

b **Define the term 'in vitro fertilisation'.** *1 mark*

  ....

c **Name the type of cell division involved in the zygote developing into an embryo.** *1 mark*

  ....

d   Following in vitro fertilisation, two embryos are often placed into the uterus of the female undergoing IVF.

    i   State why two embryos (rather than one) are often placed in the uterus.

    ii   Describe the next stage after embryos are placed into the uterus.

Contraception

a   Describe fully how implants prevent pregnancy.

b   Sterilisation is another method of contraception.

    i   Describe how female sterilisation prevents pregnancy.

    ii   Give *one* advantage and *one* disadvantage of female sterilisation.

Advantage

Disadvantage

Exam-style questions

1 **a** The diagram below represents a foetus in the uterus.

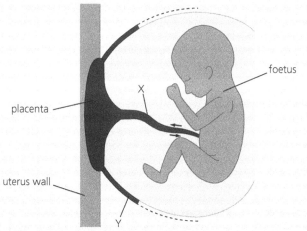

placenta

X

foetus

uterus wall

Y

   i Name structures X and Y. `2 marks`

      X ....

      Y ....

   ii On the diagram, use the letter Z to label the amniotic fluid. `1 mark`

   iii Describe fully the function of structure Y. `2 marks`

      ....

      ....

   iv Name *two* substances that pass from the placenta to the foetus. `2 marks`

      1 ....

      2 ....

b The placenta is adapted by having villi that extend into the wall of the uterus. Explain the function of the villi. `2 marks`

....

....

....

c Shortly before birth the 'waters break'. In terms of structures within the uterus, suggest what is meant by this phrase. `2 marks`

....

....

Questions marked **1** **1** are for Science Double Award and GCSE Biology students

2 The diagram below shows how the thickness of the lining of the uterus changes during the menstrual cycle.

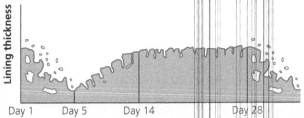

a Use the diagram to describe how the uterine lining changes between day 1 and day 14. **2 marks**

....

....

....

b Suggest when the level of oestrogen will be at its highest. Explain your answer. **2 marks**

....

....

c i Name the hormone mainly responsible for maintaining a thick uterine lining after day 14. **1 mark**

....

ii Explain why it is important that the uterus has a thick lining at that time. **2 marks**

....

....

d The diagram represents the lining of the uterus at a time when pregnancy has not taken place. What is the evidence for this? **1 mark**

....

Questions marked ❶ ❶ are for GCSE Biology students only

Genome, chromosomes, DNA and genetics

DNA is the molecule in chromosomes and genes that determines how living organisms develop. Cell division (mitosis and meiosis) explains how the chromosomes in a cell pass into daughter cells and why the chromosome number of a species remains constant. Genetics explains how characteristics in parents pass to their offspring. Genetics also explains how inherited diseases run through families. Genetic screening involves analysing the DNA in individuals to see if they (or their children) are likely to suffer from a genetic condition. Genetic engineering is a process that modifies the genome of an organism to introduce desirable characteristics.

The genome, chromosomes, genes and DNA

1 **Complete the following sentences.**   *3 marks*

All the DNA in an organism is called its .... The DNA is found in long thread-like

structures called ...., which are subdivided into smaller coding units called

....

2 a i **Name the *three* components of DNA.**   *3 marks*

....

ii **What name is given to the structure of DNA?**   *1 mark*

....

b **Explain what is meant by the base triplet hypothesis.**   *3 marks*

....

....

Cell division

3 a **Give the *three* functions of mitosis.**   *3 marks*

1 ....

2 ....

3 ....

b **The table below summarises the processes of mitosis and meiosis. Complete the table.**   *3 marks*

	Type of cell division	
	Mitosis	Meiosis
Where it takes place	Throughout the body	
Comparison with parental cell		Daughter cells have half the number of chromosomes
Comparison with other daughter cells	Same number of chromosomes and the chromosomes are the same	

Questions marked **1** **1** are for Science Double Award and GCSE Biology students

Genetic diagrams, terminology and genetic conditions

4 a Explain the difference in each of the following two sets of terms:

    i dominant and recessive alleles

    ii homozygous and heterozygous genotypes

b In terms of homozygous dominant, homozygous recessive and/or heterozygous, state the combination of parents that will give a:

    i 3:1 ratio in offspring

    ii 1:1 ratio in offspring

5 Describe and explain the process and function of a test cross.

6 Cystic fibrosis, Huntington's disease, haemophilia and Down's syndrome are examples of genetic conditions. Which of these conditions:

a is not inherited (passed down from parent to child)?

b is caused by the presence of a dominant allele?

c is caused by the presence of an extra chromosome?

d is sex linked?

Genetic screening and genetic engineering

7 a i Describe what is meant by the term 'genetic screening'.

    ii Give *one* disadvantage of making genetic information public.

b Amniocentesis and a blood test can both be used to check if a foetus is likely to be born with Down's syndrome. Give *one* advantage and *one* disadvantage of amniocentesis compared with the blood test.

Advantage

Disadvantage

Questions marked **5** **7** are for GCSE Biology students only

Exam-style questions

1 **a** Describe the structure and function of DNA.

In this question you will be assessed on your written communication skills, including the use of specialist scientific terms.

(6 marks)

....

....

....

....

....

....

....

....

....

....

....

b It is important that the number of chromosomes (and the amount of DNA) remains constant in a species over time. Explain how meiosis ensures that chromosome numbers do not change over time.

(3 marks)

....

....

....

....

2 **a** In terms of X and Y chromosomes, explain why there are approximately equal numbers of human males and females born.

(3 marks)

....

....

....

b Red-green colour blindness is a condition where affected people are unable to distinguish between different shades of red and green. It is caused by a recessive allele on the X chromosome. (**B** = normal; **b** = colour blind)

i State the phenotype shown by the genotype $X^B X^b$.

(2 marks)

....

....

....

Questions marked **1** **1** are for Science Double Award and GCSE Biology students

ii Complete the Punnett square below to show the genotypes and phenotypes in offspring arising from a cross between a parent of genotype X^BX^b and a normal male. You should also show the ratio of phenotypes produced.

4 marks

10

3 a Define the term 'genetic engineering'. **1 mark**

....

b Many people who have diabetes need to take insulin as part of their medication. Before the development of genetic engineering, they had to use insulin extracted from cattle and pigs in abattoirs.

i State the role of insulin in the body. **1 mark**

....

....

ii Give *two* advantages of using human insulin rather than insulin from domestic animals. **2 marks**

1 ....

2 ....

c As part of the process of producing insulin by genetic engineering, the human gene for insulin has to be removed from human cells and inserted into a bacterial plasmid. The removal of the human gene and the cutting of the plasmid are done using special enzymes.

i Give the name of these enzymes. **1 mark**

....

The diagram below shows how these enzymes cut the human gene and the plasmid to leave overlapping DNA strands.

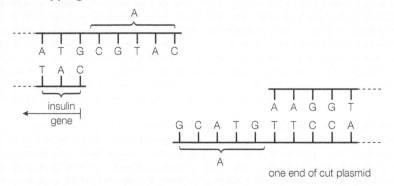

one end of cut plasmid

ii State the name of the sections labelled A.

....

iii Explain fully the reason for cutting the DNA in such a way as to leave these sections.

....

....

d Following the manufacture of the insulin, during downstreaming, 'extraction' and 'purification' are required. Explain what is meant by these terms. `2 marks`

Extraction

....

....

Purification

....

....

Variation and natural selection

Living organisms show both continuous and discontinuous variation. This variation will be due to genetic and/or environmental factors. Natural selection describes the process in which the variations within organisms can increase (or decrease) their ability to survive, reproduce and pass their genes on to the next generation. Natural selection can lead to evolution. Selective breeding describes the process in which people (not nature) favour particular features in food crops and domestic animals and control their breeding to ensure that all (or a higher proportion) have the desired features.

Types of variation

1 a Complete the table below about types of variation.

Type of variation	Example	How normally represented in graphs
Continuous		Histogram
	Tongue rolling	

b Complete the sentences below.

Genetic variation in living organisms can be caused by sexual reproduction or by

.... The latter involve random changes in the number of ....

or the structure of a .... As well as genetic differences, variation can also be

caused by the ....

Natural selection

2 a i The process of natural selection can be summarised by the statements below, but they are not in the correct sequence.

   1 The best-adapted phenotypes reproduce and pass on their genes.

   2 Phenotypes in a population vary.

   3 Due to competition, the best-adapted phenotypes survive and the less well adapted die out.

   Place the statement numbers in the correct sequence to best describe the process of natural selection.

   ....

   ii Give *one* example of natural selection.

   ....

   b i State the *two* possible outcomes of evolution on species.

1 ....

2 ....

   ii Explain fully how fossils can provide very strong evidence for evolution. **2 marks**

   ....

   ....

c   Name the term that describes a species dying out over time.   1 mark

....

Selective breeding

3 Give *two* characteristics that breeders might selectively breed into domestic animals.   2 marks

1 ....

2 ....

Exam-style questions

1 **a**   The bar chart below shows information on the blood groups of a number of blood donors.

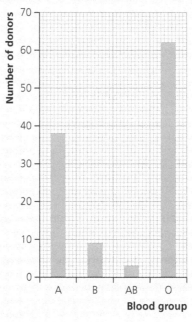

i   Calculate the difference between the number of donors of blood group O and those of blood group AB.   2 marks

Show your working.

....

ii   Name the *type* of variation shown by the data in the graph.   1 mark

....

iii   Give *one* other example of this type of variation.   1 mark

....

iv   Name the *cause* of this type of variation.   1 mark

....

Questions marked   are for Science Double Award and GCSE Biology students b Explain fully the type and causes of variation of height in humans. `3 marks`

....

....

2 a Wild garlic plants produce a toxin in their leaves that slugs find distasteful. An investigation compared the concentration of this toxin in garlic leaves in plants on a small offshore island with the main British populations. It was found that the garlic plants in mainland Britain had much more toxin than those on the offshore island.

Further investigation showed that there were no slugs on the offshore island, unlike mainland Britain, where they are very common.

i In terms of natural selection, explain how and why the garlic plants in the mainland populations have developed high levels of toxin in their leaves. `5 marks`

....

....

....

....

....

....

ii Suggest what would initially happen to the plants on the offshore island if people introduced slugs there. Explain your answer. `2 marks`

....

....

b Over thousands of years people have cultivated wheat with the result that modern wheat plants are much less variable in yield than ancient wheat.

i Which graph line (X or Y) represents modern wheat? Explain your answer. `1 mark`

....

....

ii Describe how selective breeding has produced the change shown in the graph. `3 marks`

....

....

....

iii Give *one* difference between selective breeding and natural selection. `1 mark`

....

Health, disease, defence mechanisms and treatments

Our body uses a number of strategies to defend against communicable (infectious) and non-communicable (non-infectious) diseases. Medicines, including vaccinations and antibiotics, are also important in helping us avoid or recover from disease. Many non-communicable diseases, such as heart disease, strokes and some cancers, are linked to lifestyle. We can reduce our chances of getting these diseases by avoiding those lifestyle factors that increase the risk of developing these diseases.

Communicable diseases and the body's defence mechanism

1 a **What is meant by the term 'communicable disease'?** `1 mark`

....

b **The table below shows features of some communicable diseases caused by bacteria. Complete the table.** `4 marks`

Name of disease	How spread	Prevention	Treatment
Chlamydia		Using a condom	
Salmonella	Contaminated food		Antibiotics
Tuberculosis		Vaccination	Antibiotics (and other drugs)

2 a **Describe how mucous membranes in the nose can prevent infection.** `2 marks`

....

....

....

b i **State precisely what causes the body to produce antibodies?** `1 mark`

....

ii **Name the type of white blood cell that produces antibodies.** `1 mark`

....

iii **Describe fully how antibodies combat disease.** `3 marks`

....

....

....

....

Development of medicines, antibiotics and vaccinations

3 Complete the sentences below about the discovery and development of penicillin.   **3 marks**

The scientist .... discovered that the bacteria on an agar plate were

prevented from growing by a substance diffusing from a fungus that had contaminated the agar

plate. The chemical responsible for preventing the growth of bacteria was later isolated by two

other scientists called .... and ....

4 The table below shows some of the stages in the development of medicines and some of the reasons for each stage. Complete the table.   **3 marks**

Stage	Description of stage	Reason for stage
Preclinical trials — in vitro (in lab)		Checks that the drug is effective and not poisonous before testing on living organisms
Preclinical trials — animal testing	Testing on animals (living organisms)	
	Testing on human volunteers (and patients)	Finding the optimum dosage

5 a i What are superbugs?

....

ii Give an example of a superbug.

....

b Superbugs have been a particular problem in some hospitals. State *two* measures that can be taken to reduce the number of superbugs in hospitals.

1 ....

2 ....

6 a Explain why the microorganisms in vaccinations must be modified (or killed).

....

b The graph below shows how the antibody level in the body changes after initial and follow-up (booster) vaccinations.

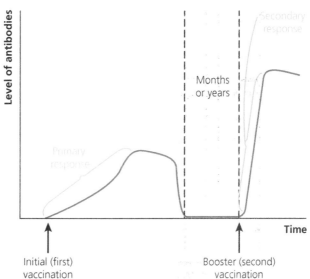

 i The graph shows that the secondary response produces antibodies faster than the primary response. Give *one* other difference. 1 mark

 ii Explain why the booster vaccination resulted in antibodies being produced faster. 2 marks

c Vaccinations produce active immunity. State what is meant by active immunity. 1 mark

Non-communicable diseases

7 Lifestyle factors can contribute to non-communicable diseases.

 a i Name the disease that can be caused by overexposure to the sun. 1 mark

 ii Explain how too much sunlight can cause this disease. 2 marks

 b Smoking is another lifestyle choice that can cause harm. Name the chemical in cigarette smoke that is addictive. 1 mark

8 a The statements below explain how a heart attack occurs, but they are not in the correct sequence.

  1 The cells in the heart muscle die.

  2 A build-up of cholesterol leads to a blockage in a coronary artery.

  3 Not enough oxygen or glucose reaches the heart cells.

  4 The heart cells cannot respire to produce energy.

  Place the statement numbers in the correct sequence to describe how a heart attack occurs. 2 marks

 b Explain the role of angioplasty in the treatment of heart disease. 2 marks

9 a State *two* differences between malignant and benign tumours. 2 marks

  1

  2

b Surgery is one method used to treat cancer.

i Give *one* advantage and *one* disadvantage of surgery as a treatment for cancer. **2 marks**

Advantage ....

Disadvantage ....

ii Give *one* other method used to treat cancer. **1 mark**

....

Aseptic techniques

10 a State the method used to sterilise Petri dishes and culture media. **1 mark**

....

b Explain why Petri dishes are 'taped' once they have had microorganisms added to them. **1 mark**

....

c What term is used to describe the transfer of microorganisms from a culture bottle to a sterile agar plate. **1 mark**

....

Exam-style questions

11

1 a The flu is a communicable disease caused by microorganisms.

i Name the type of microorganism that causes flu. **1 mark**

....

ii Describe how the flu is spread from person to person. **1 mark**

....

b There are many types of flu strain in any one year. In each strain of flu, the microorganism responsible is slightly different. In terms of antibody–antigen reactions, explain why someone could be immune to one strain of flu but not another. **3 marks**

....

....

....

....

c Antibiotics are chemicals used to treat disease.

i Explain why antibiotics are not prescribed for people who have flu. **2 marks**

....

....

Questions marked **1** **1** are for GCSE Biology students only

ii Explain how and why antibiotic resistance develops in bacteria. Your answer should include the changes that occur to bacteria to make them resistant. **4 marks**

....

....

....

....

2 a i State what is meant by the term 'non-communicable disease'. **1 mark** ⓰

....

ii State the *two* main causes of non-communicable diseases. **2 marks**

....

b Smoking can cause lung cancer and other non-communicable diseases.

i Apart from lung cancer, name *two* other non-communicable diseases caused by smoking. **2 marks**

....

ii People who smoke often are very short of energy. Taking account of the various effects of smoking, as appropriate, explain why a smoker can be short of energy. **5 marks**

....

....

....

....

....

c Four hundred newly diagnosed patients took part in a clinical trial on a new drug to treat lung cancer. Two hundred people were given the drug and the other 200 were given a placebo — a placebo looks the same as the drug but has no active ingredient.

The trial started in 2010 and ran for 5 years. The numbers of patients surviving are shown in the table below.

| | Patients still alive | |
Year	Taking the drug	Taking the placebo
2010	200	200
2011	188	161
2012	181	120
2013	83	82
2014	45	44
2015	4	6

i Calculate the total percentage of patients still alive at the end of the trial. **2 marks**

Show your working.

.... %

Questions marked ① ① are for Science Double Award and GCSE Biology students

ii Using data from the table, describe the effectiveness of the trial drug. `3 marks`

....

....

....

iii State the function of the placebo. `1 mark`

....

---

3 **a** Aseptic techniques refer to procedures used to transfer microorganisms without causing contamination.

A student was transferring bacteria from a culture bottle to a Petri dish. She used the flame of a Bunsen burner to sterilise an inoculating loop. Immediately after flaming the inoculating loop, she placed it into the culture bottle to obtain some of the bacteria to be transferred. She then removed the Petri dish lid and set it down carefully on a clean part of the laboratory bench. At this stage, she transferred the bacteria to the Petri dish by gently gliding the loop over the agar. After this was done, she then placed the Petri dish lid back on and placed the loop and the culture bottle in the sink for washing.

This student did not show good aseptic technique when carrying out the transfer. Using the information provided, state *three* errors that she made. `3 marks`

1 ....

....

2 ....

....

3 ....

b Many plants produce antimicrobial chemicals that kill fungi and bacteria that could cause them harm. To investigate the antimicrobial properties of two species of plant, for example wild garlic and mint, you are provided with leaves from each species, small absorbent discs of filter paper, sterile Petri dishes containing agar and any other items of apparatus normally found in a laboratory.

i Describe how you could set up an investigation to compare the antimicrobial properties of the two species. (You do not need to describe any aseptic techniques in your answer.) `4 marks`

....

....

....

....

ii Give *two* variables that should be controlled in this investigation. `2 marks`

1 ....

2 ....

Questions marked 🅑 🅑 are for GCSE Biology students only

iii Apart from producing antimicrobial chemicals, name *one* other way in which plants are adapted to defend against disease.

1 mark

....

4 a Describe and explain the link between early detection and survival in cancer patients.

2 marks

....

....

b Radiotherapy involves killing cancer cells using X-ray or gamma radiation. The diagram below shows how the source of the radiation changes during radiotherapy.

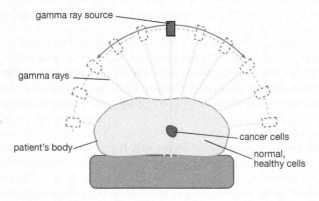

gamma ray source

gamma rays

patient's body

cancer cells

normal, healthy cells

Use the diagram and your knowledge to explain fully why the position of the radiation source changes during radiotherapy.

2 marks

....

....

c Explain how immunotherapy can help treat cancer.

3 marks

....

....

Hodder Education, an Hachette UK company, Carmelite House, 50 Victoria Embankment, London EC4Y 0DZ

Orders

Hachette UK Distribution, Hely Hutchinson Centre, Milton Road, Didcot, Oxfordshire, OX11 7HH

tel: 01235 827827

e-mail: education@hachette.co.uk

Lines are open 9.00 a.m.–5.00 p.m., Monday to Friday.

You can also order through the Hodder Education website: www.hoddereducation.co.uk

© James Napier 2018

ISBN 978-1-5104-1908-7

First printed 2018

Impression number 8

Year 2023

This guide has been written specifically to support students preparing for the CCEA GCSE Biology examinations. The content has been neither approved nor endorsed by CCEA and remains the sole responsibility of the author.

Typeset by Aptara, India

Printed in Great Britain by Ashford Colour Press Ltd

Hachette UK's policy is to use papers that are natural, renewable and recyclable products and made from wood grown in well-managed forests and other controlled sources. The logging and manufacturing processes are expected to conform to the environmental regulations of the country of origin.

HODDER EDUCATION

t: 01235 827827

e: education@hachette.co.uk

w: hoddereducation.co.uk

ISBN 978-1-5104-1908-7

MIX
Paper | Supporting responsible forestry
FSC™ C104740